l Oberleutnant Immelmann
rer Kameraden.

373
Postkartenvertrieb W. S
BERLIN N. 37.

World War I German Aviators

The Sanke Cards

Also by the Author

Uniforms & Equipment of the Imperial German Army 1900-1918 (two volumes)

The Kaiser's Army In Color: Uniforms of the Imperial German Army as Illustrated by Carl Becker 1890-1910

German Uniforms, Insignia & Equipment 1918-1923: Freikorps, Reichswehr, etc.

First to the Front: The Aerial Adventures of 1st Lt. Waldo Heinrichs and the 95th Aero Squadron 1917-1918

The Hat in the Ring Gang: The Combat History of the 94th Aero Squadron in World War I

World War I German Aviators

The Sanke Cards

Charles Woolley

Schiffer Military History
Atglen, PA

Book design by Robert Biondi.

Copyright © 2003 by Charles Woolley.
Library of Congress Catalog Number: 2003103334.

All rights reserved. No part of this work may be reproduced or used in any forms or by any means – graphic, electronic or mechanical, including photocopying or information storage and retrieval systems – without written permission from the publisher.

The scanning, uploading and distribution of this book or any part thereof via the Internet or via any other means without the permission of the publisher is illegal and punishable by law. Please purchase only authorized editions and do not participate in or encourage the electronic piracy of copyrighted materials.

"Schiffer," "Schiffer Publishing Ltd. & Design," and the "Design of pen and ink well" are registered trademarks of Schiffer Publishing, Ltd.

Printed in China.
ISBN: 0-7643-1879-9

We are always looking for people to write books on new and related subjects. If you have an idea for a book, please contact us at the address below.

Published by Schiffer Publishing Ltd.
4880 Lower Valley Road
Atglen, PA 19310
Phone: (610) 593-1777
FAX: (610) 593-2002
E-mail: Info@schifferbooks.com.
Visit our web site at: www.schifferbooks.com
Please write for a free catalog.
This book may be purchased from the publisher.
Please include $3.95 postage.
Try your bookstore first.

In Europe, Schiffer books are distributed by:
Bushwood Books
6 Marksbury Ave.
Kew Gardens
Surrey TW9 4JF
England
Phone: 44 (0)20 8392-8585
FAX: 44 (0)20 8392-9876
E-mail: Bushwd@aol.com.
Free postage in the UK. Europe: air mail at cost.
Try your bookstore first.

Introduction

Hungary in 1869 witnessed the birth of the first postcard, a single sheet of stiff paper with an illustration on one side and space for a brief message and an address on the other. This rather simple invention was the brainchild of a Doctor Emanuel Herrmann. The Hungarian postal authority gave its approval for the use of this type of written communication that year, and other countries were quick to follow.

The first commercially published postcards were made available to the opposing forces of Germany and France in 1870 during the Franco-Prussian War. The concept of these cards was particularly attractive to soldiers fighting at the front where supplies of writing paper and envelopes necessary for letters home were scarce as well as expensive. At the end of hostilities, the use of postcards was fully entrenched with cards produced in Germany at the forefront of popularity.

In the second half of the nineteenth century, the Germans were particularly advanced in the art of color printing. The process of chromolithography, color printing in which the final brightly hued image was produced from an engraved series of stone or zinc plates. Each plate during the printing process carried a different colored ink that allowed for the artwork to be portrayed in multiple rich colors. Due to the skill of German chromolithographers and the reasonable rates they charged, the majority of postcards printed in the late 1800s and early 1900s, worldwide, bore the legend "Printed in Germany."

The brilliant colors natural to military uniforms worn during the time caused artists in England, France and Germany to turn their hands to the creation of many series of richly colored cards depicting military subjects. In England, Harry Paine, John McNeill, Edgar Holliday, and Ernest Ibbetson painted uniform cards that were published and sold by the firms of Gale and Polden Ltd. of Aldershot and G. Falkner and Sons of London. In Germany, Anton Hoffmann, Oscar Merté, Carl Becker, and H. Lüders painted military subjects for the many firms who distributed cards throughout the country and abroad. Not only did these artists paint single or pairs of uniformed figures, but, also, they painted views of military life often depicted in an almost cartoon-like humorous vein. Soldiers who were on active duty performing their required military service were the best consumers of these cards and bought them in quantity to send best wishes home and to friends.

At the same time enterprising photographers and their studios were hard at work taking photographs of uniformed men in service which were printed in postcard format making it easy to send home to family and loved ones. The photographers had studios in towns near the training Kasernes and often would set up their cameras in portable studios on the bases themselves. Companies producing and distributing "real photo" postcards, NPG, Gustav Liersch and Company and W. Niederastroth were at the same time offering fine portrait views of the Kaiser, his family and the more famous generals resplendent in their full dress uniforms. At this point, due to their vast variety and attractive nature of these subjects, customers began to purchase these handsome cards forming collections which could be mounted in specially designed albums rather than use them for their original intended communication use. This trend in the market caused the distributors to turn to additional subjects to further capture the public's eye, naval ships, the Kaiser's maneuvers, parades, and to one special area, the craze about manned flight.

In 1913 the hub of German aerial activity was the vast airfield at Johannisthal where lighter than airships were housed and where those daring young men, and yes, women, were airborne in their flying machines. These activities captured the hearts of the public and the fertile marketing imagination of one Willi Sanke. His advertisements soon began to appear declaring the availability, for only ten pfennigs each, of "Offizielle Ansichtkarten" (official picture postcards) of the Flugplatz Johannisthal which included not only the flyers, but also their flying machines and the colossal airships. In his early cards, Sanke utilized many photographs taken by one Franz Fisher and Sanke cards carried his name and copyright up to at least card number 271. Card number 310 was the next number encountered and Fisher's name and copyright were no longer in evidence thereafter. Names of other photographers are credited on many of Sanke's cards. Professor Nicola Perschied of Berlin, Professor Krath, Adolph Jensen, C.J. von Duhren, Carl Selberg, E.M. Kollstede, A. Buck of München, and Hugo Erfurth were all given credits.

The photographers responsible for the majority of Sanke's cards remain anonymous and possibly were his employees, commanding no credit line or copyright.

An early advertisement announced his firm offering "about 200 views" of the Flug-platzes Johannisthal, his company was called "Postkarten-Vertreib Willi Sanke" and was located at Schönhauser Allee 185, Berlin N.37. In Sanke's early offerings Count Zeppelin's airships were the subjects of cards numbered two through five and ten through fourteen, with the first heavier than air Rumpler Taube bearing card number 73. Sanke's numbering system appears to have been somewhat erratic, as large blocks of sequential numbers have no yet to be identified cards, while there are many known cards with no numbers whatsoever. The subject matters which evolved included pre-War Zeppelin-Luftschiffs, early flyers and aircraft, and after 1914, military and naval aircraft, decorated military and naval pilots and observers, and a non-aviation subject, Captain Nerger, the famed Imperial Naval commander of the sea-raider "Wolf." Throughout the production of Sanke cards, many carried a small logo, a circle surrounding a large "R", a smaller "P" through the upper part of the "R" and an even smaller "H" was placed between the legs of the "R". The meaning of the letters and their placement remains a mystery to the author.

When card number 340 and card number 363, splendid pictures of Max Immelmann and of Oswald Boelcke were issued, every schoolboy in Germany rushed to purchase his own copy. An immediate new and popular market was born, for which Sanke hurried to issue more cards of additional "Unser erfolgreicher Kampfflieger," our successful war flyers. The tremendously enthusiastic response to the images of these heroic men of the air can be, in some way, compared to the popularity of cards depicting famous baseball players issued in the United States and the craze to collect them. Copies of Sanke cards were sent by many eager German youths to the individual flyers requesting that their autographs be added and returned to the sender. Many of the pilots themselves maintained a supply to respond to autograph requests not accompanied by a card or to pass on to friends as a keepsake, similar to the use of the *carte-de-visite* of the late nineteenth century.

If a particularly famous and well-known flyer were to meet his untimely death, memorial editions of existing cards were quickly issued; some cards were simply bordered in black and some made available with heroic and sentimental verse added below the flyer's picture. Of the eleven known Manfred von Richthofen cards sold by Sanke, three were memorial in nature.

The interest by collectors today in Sanke cards is no less intense than that of the schoolboys and adults of Germany at the time they were being sold during the Great War. Values have risen steadily over the past eighty plus years and the cards which first were offered for 10 pfennigs, are now being sold for sums over a hundred dollars, particularly those bearing the subject's autograph.

For this book, every effort has been made to include each of the personality cards produced by Willi Sanke, but it must be noted that there is the possibility of an extremely rare card having escaped the dragnet of the author's and other's efforts. Information regarding the details of any other known cards omitted would be greatly appreciated, forwarded to me through the publisher. There are many fine books available today which give detailed accounts of the lives and combat performances of the men photographed in Sanke's cards, as well as of many other German flyers who were not featured in his offerings.

This book is about Sanke cards depicting the pilots, gunners and observers and how they looked. It is not intended to be a group of in depth biographies of the famous flyers and observers of the Imperial German Air Service. Therefore, a bibliography is included for those who wish to delve into that facet of the subject in greater detail.

The assembly of the pictured cards in this volume would have been impossible without the assistance of many. Dr. Erik Carlson, who made available the cards in the History of Aviation Collection at the University of Texas, Dallas, as well as the onsite help of Ken Rice, member of the History of Aviation's volunteer staff. The superb collections of Mike Stacey, Dr. Marc Magness, and Jeffrey Sands were generously shared and filled in those blanks not covered by the University of Texas, Dallas holdings originally amassed by the late Ed Ferko. These collectors are passionate about their subject, always in search of those last few cards to complete their collections, or the elusive card not known before. To these men, I dedicate this book. My old friend Pete Grosz led me to the work done by Günter Otto and Bernd Stasche of Germany in compiling their list of Sanke cards to which we now have added a few. Thanks to all.

The Cards

Adam, Hans Ritter von, *Leutnant*
Born Eisenstein, 24 May 1886
Knight's Cross with Swords of the
Royal Hohenzollern House Order
Sanke 557

FAA 2 b, JASTA 34, 6
Killed in Action, 15 November 1917
21 Victories
2 February 1918

Allmenröder, Karl, *Leutnant*
Born Wald, 3 May 1896
Pour le Mérite, 14 June 1917
Sanke 543

FA 227, JASTA 11
Killed in Action, 27 June 1917
30 Victories

Althaus, Ernst Freiherr von, *Oberleutant*
Born Coburg, 19 March 1890
Pour le Mérite, 21 July 1916
Sanke 382, 383, 430, 431

FA 23, KEK VAUX, JASTA 4, 14, 10
Died, 29 November 1946
9 Victories

Unser erfolgreicher
Kampfflieger
Oberleutnant
Freiherr von Althaus.
383
Postkartenvertrieb W. Sanke
BERLIN N. 37.
Nachdruck wird gerichtlich verfolgt.

Unser erfolgreicher
Kampfflieger
Oberleutnant
Freiherr von Althaus.
430
Postkartenvertrieb W. Sanke
BERLIN N. 37
Nachdruck wird gerichtlich verfolgt.

Unser erfolgreicher Kampf-Flieger
Oberleutnant
Freiherr von Althaus.
431
Postkartenvertrieb W. Sanke
BERLIN N. 37.
Nachdruck wird gerichtlich verfolgt.

Anslinger, Leopold, *Leutnant* FA 54, FA 24, KEST 9
Born Freiburg, 13 August 1891 Died, 18 June 1978
Ritterkreutz of the Royal Hohenzollern House Order 10 Victories
Sanke 565

Baeumer, Paul, *Leutnant* (Bäumer)
Born Duisberg-Ruhrart, 11 May 1896
Pour le Mérite, 2 November 1918
Sanke 630

FA 7, JASTA 2, 5
Died in test crash, 15 July 1927
43 Victories

Baldamus, Hartmut, *Leutnant*
Born Dresden, 10 August 1891
Knight's Cross with Swords of the Royal Hohenzollern House Order, 8 January 1917
Sanke 390, 432

FFA 20, JASTA 5, 9
Killed in Action, 14 April 1917
18 Victories

Unser erfolgreicher
Kampfflieger
Leutnant Baldamus
432
Postkartenvertrieb W. Sanke
BERLIN N. 37.
Nachdruck wird gerichtlich verfolgt.

Bernert, Fritz Otto, *Oberleutnant*
Born Ratibor, Upper Silesia, 6 March 1893
Pour le Mérite, 23 April 1917
Sanke 442, 443, 521

FFA 27, 71, KEK VAUX, JASTA 4, 2, 6
Died Influenza, 18 October 1918
27 Victories

Unser erfolgreicher Kampf-Flieger
Leutnant Bernert
443
Postkartenvertrieb W. Sanke
BERLIN N. 37.
Nachdruck wird gerichtlich verfolgt.

Unser erfolgreicher Kampf-Flieger
Leutnant Bernert
521
Postkartenvertrieb W Sanke
BERLIN N. 37
Nachdruck wird gerichtlich verfolgt

Berr, Hans, *Oberleutant*
Born Braunschweig, 20 May 1890
Pour le Mérite, 4 December 1916
Sanke 425, 426

KEK AVILLERS, JASTA 5
Killed in Action, 6 April 1917
10 Victories

Unser erfolgreicher
Kampf-Flieger
Oberleutnant Berr.
426
Postkartenvertrieb W. Sanke
BERLIN N. 37
Nachdruck wird gerichtlich verfolgt.

Berthold, Rudolf, *Hauptmann*
Born Ditterswind, 24 March 1891
Pour le Mérite, 12 October 1917
Sanke 402, 423, 629

FAA 23, KEK VAUX, JASTA 4, 14, 18, JG II
Killed by Communists, 15 March 1920
44 Victories

Unser erfolgreicher Kampf-Flieger
Oberleutnant Berthold.
423
Postkartenvertrieb W. Sanke
BERLIN N. 37.
Nachdruck wird gerichtlich verfolgt.

Unser erfolgreicher Kampf-Flieger
Hauptmann Berthold
629
Postkartenvertrieb W. Sanke
BERLIN N. 37.
Nachdruck wird gerichtlich verfolgt

Bertrab, Joachim von, *Leutnant*	JASTA 30
Prisoner of War, 12 August 1917	
Iron Cross First Class	5 Victories
Sanke 516, 530, 531	

Leutnant v. Bertrab
530
BERLIN N 3
Phot. Nicola Perscheid

Leutnant v. Bertrab
531
Postkartenvertrieb W. Sanke
BERLIN N. 37
Phot. Nicola Perscheid

Unser erfolgreicher Kampfflieger
Leutnant Blume

Blume, Walter, *Leutnant*	FA 65, FA (A) 280, JASTA 26, 9
Born Hirschberg, Silesia, 10 January 1896	Died, 27 May 1964
Pour le Mérite, 2 October 1918	28 Victories
Sanke 675	

Boelcke, Oswald, *Hauptmann*
Born Giebichstein, 19 May 1891
Pour le Mérite, 12 January 1916
Sanke 363, 363, 363, 369, 393, 409

FA 13, 62, JASTA 2
Killed in Collision, 28 October 1916
40 Victories

Hauptmann Boelcke. †
363
Postkartenvertrieb W Sanke
BERLIN N. 37.
Nachdruck wird gerichtlich verfolgt.

Hauptmann Boelcke.
369
Postkartenvertrieb W. Sanke
BERLIN N. 37.
Nachdruck wird gerichtlich verfolgt.

Hoch im glitzernden Gewölke
Thront die Kühnheit! Unser Bölke!

Hoch im Aether und Azur
Kreuzt er adlergleich die Flur!
Feinde faßt ein jäh Entsetzen
Wenn sie Falkenreich verletzen!
Unten — graue Nebelfluten,
Über ihm — der Sonne Gluten,
Vorne — geisterhaft und bleich
Stürzt der Feind ins Todesreich!

Kampfgewitter schaurig schön! —
Sieg auf Sieg in Götterhöh'n!
Was, nur Sternenaugen nah,
Je an Tapferkeit geschah
Zwingt selbst Feinde voller Grollen
Dir Bewunderung zu zollen!
König der Lüfte halte Wacht
Deutsche Ehre, Deutscher Macht!

Hoch im glitzernden Gewölke
Thront die Kühnheit! Unser Bölke

(Ruth Barbara Grede.)

393
Postkartenvertrieb W. Sanke
Berlin N 37
Nachdruck wird gerichtlich verfolgt.

Fliegerhauptmann Boelcke †

O stolzer König im Fliegerreich
Luftkönig Boelcke, wer steht dir gleich?
Du höchste Tapferkeit zogst hinaus,
Um täglich Neues zu wagen,
Du hast durch Sturm und Wettergebraus
Des Vaterland's Ehre getragen!

Als strahlendster Stern am Firmament,
Den jeder Deutsche liebt und kennt,
Barg deine Jugend den Heldenmut,
Der Wunder über Wunder tut.
Durch Donner und Blitze, in Nacht und Licht
Dein Siegesstrahl sich Bahnen bricht!

Und schmerzgebeugt schaun wir hinein
In deines Ruhmes hellen Schein!
Einst wird die Nachwelt jubelnd lesen,
Was du dem Vaterland gewesen!
An Leuchtkraft steht dir keiner gleich,
Luftkönig Boelcke, im Sternenreich!

Ruth, Barbara Goede.

409
Postkartenvertrieb W. Sanke
BERLIN N. 37
Nachdruck wird gerichtlich verfolgt.

Boenigk, Oskar Freiherr von, *Oberleutnant*
Born Siegersdorf, Silesia, 25 August 1893
Pour le Mérite, 25 October 1918
Sanke 673

KASTA 19, 32, JASTA 4, 21, JG II
Died, 30 January 1946
26 Victories

Boenische, Erich, *Leutnant zur See*	1 SEAPLANE STN
Iron Cross First Class	5 Victories
Led Bombing Raid on Dover	
Sanke 412	

Böhme, Erwin, *Leutnant*
Born Holzminden, 29 July 1879
Pour le Mérite, 24 November 1917
Sanke 502

KASTA 10, JASTA 2, 29
Killed in Action, 29 November 1917
24 Victories

Bolle, Karl, *Rittmeister*
Born Berlin, 20 June 1893
Pour le Mérite, 28 August 1918
Sanke 657, 685

KG 4, KASTA 23, JASTA 28, 2
Died, 9 October 1955
36 Victories

Unser erfolgreicher Kampfflieger
Leutnant Karl Bolle
685
rtenvertrieb W. Sanke
ERLIN N 37.
wird gerichtlich verfolgt
Nicola Perscheid Berlin W. 9 phot

Bongartz, Heinrich, *Leutnant*
Born Gelsenkirchen, 31 January 1892
Pour le Mérite, 23 December 1917
Sanke 571, 603, 605

FA 5, KASTA 27, JASTA 36
Died, 23 January 1946
33 Victories

Unser erfolgreicher Kampfflieger
Leutnant Bongartz
603
Postkartenvertrieb W. Sanke
BERLIN N. 37.
Nachdruck wird gerichtlich verfolgt.

Unser erfolgreicher Kampfflieger
Leutnant Bongartz
605
Postkartenvertrieb W. Sanke
BERLIN N 37
chdruck wird gerichtlich verfolgt.

Brandenburg, Ernst, *Hauptmann*
Born Westphalia, 4 June 1883
Pour le Mérite, 14 June 1917
Sanke 536, 541

KAGOL
Died, 1952
Led Gotha Raid over London, 13 June 1917

Hauptmann Brandenburg
541
Postkartenvertrieb W. Sanke
BERLIN N. 37.
Nachdruck wird gerichtlich verfolgt.

Brauneck, Otto, *Leutnant*
Born Sulzbach, 27 February 1896
Knight's Cross of the Royal Hohenzollern House Order, 19 January 1917
Sanke 508

FFA 69, JASTA 25, 11
Killed in Action, 26 July 1917
10 Victories

Buckler, Julius, *Leutnant*
Born Mainz, 28 March 1894
Pour le Mérite, 4 December 1917
Sanke 585, 631

FAA 209, JASTA 17
Died, 23 May 1960
36 Victories

Unser erfolgreicher Kampfflieger
Leutnant Buckler.
631
Postkartenvertrieb W. Sanke
BERLIN N. 37
Nachdruck wird gerichtlich verfolgt.

Buddecke, Hans-Joachim, *Hauptmann*
Born Berlin, 22 August 1890
Pour le Mérite, 14 April 1916
Sanke 371, 433, 434

FA 23, 6, 5, JASTA 4, 30 18
Killed in Action, 10 March 1918
13 Victories

Unser erfolgreicher Kampf-Flieger
Hauptmann Buddecke
433
Postkartenvertrieb W. Sanke
BERLIN N. 37.
Nachdruck wird gerichtlich verfolgt.

Unser erfolgreicher Kampf-Flieger
Hauptmann Buddecke
434
Postkartenvertrieb W. Sanke
BERLIN N. 37
Nachdruck wird gerichtlich verfolgt.

Bülow-Bothkamp, Walter von, *Leutnant*
Born Borby, Holstein, 24 April 1894
Pour le Mérite, 8 October 1917
Sanke 582

FA 22, 300, JASTA 18, 36, 2
Killed in Action, 6 January 1918
28 Victories

Christiansen, Friedrich, *Oberleutnantzur See*	SFL 1
Born Wyk-on-Fohr, 12 December 1879	Died, 5 December 1972
Pour le Mérite, 11 December 1917	13 Victories
Sanke 608, 609	

Unser erfolgreicher Seekampfflieger
Oberleutnant Christiansen
609
Postkartenvertrieb W. Sanke
BERLIN N. 37
Nachdruck wird gerichtlich verfolgt.

Collin, Dieter, *Leutnant*
Born Luben, 17 February 1893
Iron Cross First Class
Sanke 669

JASTA 22, 2, 56
Died of Wounds, 13 August 1918
13 Victories

Cossel, Maximilian von, *Oberleutant*
Born Waldeck, Date Unknown
Knight's Cross with Swords of the Royal Hohenzollern House Order, 5 October 1916
Sanke 401, 424

FFA 62
Prisoner of War, August 1917

424

Dannhuber, Xaver, *Leutnant* (Danhuber)
Date of Birth and Location Unknown
Knight's Cross with Swords of the
Royal Hohenzollern House Order, 29 March 1918
Sanke 615

KAGOL 6, SS 25, JASTA 26, 79
11 Victories

Donhauser, Hans Christian Friedrich, *Leutnant*
Date of Birth, and Location Unknown
Golden Military Merit Cross, 9 October 1918
Sanke 677

FA 10, JASTA 17
Killed in Air Crash, 1919
19 Victories

Dörr, Gustav (Also Doerr), *Leutnant*
Born, 5 October 1887
Golden Military Cross, 29 August 1917
Sanke 683

FA 68, 6, 257, JASTA 45
Accidentally killed, 11 December 1928
35 Victories

Dossenbach, Albert, *Leutnant*
Born St. Blasien, 5 June 1891
Pour le Mérite, 11 November 1916
(First two-seater pilot)
Sanke 416

FA 22, JASTA 2, 36, 10
Killed in Action, 3 July 1917
15 Victories

Dostler, Eduard Ritter von, *Oberleutant*
Born Pottenstein, Bavaria, 3 February 1892
Pour le Mérite, 6 August 1917
Sanke 548

SS 27, KASTA 36, JASTA 13, 34, 6
Killed in Action, 21 August 1917
26 Victories

Eisenmenger, *Leutnant, Observer* and
Gund, *Vizefeldwebel, Pilot*
Both awarded the Iron Cross First Class
Sanke 636

On Observation Mission, they were attacked by six enemy fighters, four of which Eisenmenger shot down.

Eschwege, Rudolf von, *Leutnant*
Born near Hamburg, 25 February 1895
Knight's Cross of the Military St. Henry Order, 8 July 1917
Sanke 549

FA 36, 66, 30
Killed in Action, 21 November 1917
20 Victories

Fabeck, Paul, *Oberflugmeister* Imperial German Navy
Pilot for Stein from the raider "Wolf"
Sanke 635

Fahlbusch, Wilhelm, *Leutnant*
Born Hannover, 8 February 1892

KASTA 1
Killed in Action, 6 September 1916
5 Victories

Sanke 418

Falke, *Oberleutnant, Observer*
Iron Cross First Class
Sanke 589

Festner, Sabastian, *Vizefeldwebel*
Born Holzkirchen, Bavaria, 30 June 1894
Member's Cross with Swords of the Royal Hohenzollern House Order, 23 April 1917
Sanke 510

FA 18, 5, JASTA 11
Killed in Action, 25 April 1917
12 Victories

Fokker, Anthony Herman Gerard, *Direktor*
Born Kediri, Java, 6 April 1890 Died, 23 December 1939
Sanke 359, 410, 611

Direktor Fokker
Erbauer des Fokker-Kampfflugzeuges
410
Postkartenvertrieb W. Sanke
BERLIN N. 37.
Nachdruck wird gerichtlich verfolgt.

Fokker
Direktor Fokker,
Erbauer der Fokker-Kampfflugzeuge
611
Postkartenvertrieb W. Sanke
BERLIN N. 37.
Nachdruck wird gerichtlich verfolgt.

Francke, Rudolf, *Leutnant*	SS 17, JASTA 2, 8
Golden Military Merit Cross, 16 May 1917	15 Victories
Sanke 648	

Frankl, Wilhelm, *Leutnant*
Born Hamburg, 20 December 1893
Pour le Mérite, 12 July 1916
Sanke 384, 420, 421

FA 40, KEK VAUX, JASTA 4
Killed in Action, 8 April 1917
20 Victories

Unser erfolgreichster Flieger
Leutnant Frankl.
420
Postkartenvertrieb W. Sanke
BERLIN N. 37

Unser erfolgreichster Flieger
Leutnant Frankl.
421
Postkartenvertrieb W. Sanke
BERLIN N. 37.
Nachdruck wird gerichtlich verfolgt.

Frickert, Wilhelm, *Leutnant*
Knight's Cross with Swords of the Royal Hohenzollern House Order, 5 May 1918
Sanke 598, 562, 561

FA 24/242, 20, JASTA 64, 65
12 Victories

Unser erfolgreicher Kampfflieger
Offizierstellvertreter Frickart
562
Postkartenvertrieb W. Sanke
BERLIN N. 37.
Nachdruck wird gerichtlich verfolgt.

Unser erfolgreicher Kampfflieger
Offizierstellvertreter Frickart
561
Postkartenvertrieb W. Sanke
BERLIN N.37

Fricke, Hermann, *Oberleutnant, Observer*
Born Munster, 16 June 1890
Pour le Mérite, 23 December 1917
Sanke 607

FA 2
160 Combat Missions

Friedrich, Alfred, *Vizefeldwebel* FA 1
Sanke 394, 395

Unser erfolgreicher Flieger
Alfred Friedrich
Postkartenvertrieb W. Sanke
BERLIN N. 37.
Nachdruck wird gerichtlich verfolgt.
395

Gerlich, *Oberleutnant*, Observer Adjutant KAGOHL 3
Sanke 388

Gontermann, Heinrich, *Leutnant*
Born Siegen, 25 February 1896
Pour le Mérite, 14 May 1917
Sanke 517, 527, 528, 529, 558, 559

FA 25, JASTA 5, 15
Accidentally killed, 30 October 1917
39 Victories

Leutnant
Gontermann.
527
Postkartenvertrieb W. Sanke
BERLIN N. 37.
Nachdruck wird gerichtlich verfolgt.

Leutnant
Gontermann.
Phot. Nicola Perscheid
528
Postkartenvertrieb W. Sanke
BERLIN N. 37.
Nachdruck wird gerichtlich verfolgt.

Leutnant
Gontermann.
529
Postkartenvertrieb W. Sanke
BERLIN N. 37.
Nachdruck wird gerichtlich verfolgt.

Unser erfolgreicher Kampfflieger
Leutnant Gontermann.
558
Postkartenvertrieb W. Sanke
BERLIN N. 37.
Nachdruck wird gerichtlich verfolgt.

Unser erfolgreicher Kampfflieger
Leutnant Gontermann.
559
Postkartenvertrieb W. Sanke
BERLIN N. 37.
Nachdruck wird gerichtlich verfolgt.

Göring, Herman Wilhelm, *Oberleutnant*
Born Rosenheim, 12 January 1893
Pour le Mérite, 2 June 1918
Sanke 535, 573, 654, 655

FFA 25, KEK STENAY, KEK METZ, JASTA 7, 5, 26, 27, JG 1
Died by Suicide, 15 October 1946
22 Victories

Oberleutnant Göring
573
Postkartenvertrieb W. Sanke
BERLIN N. 37
Nachdruck wird gerichtlich verfolgt.

Unser erfolgreicher Kampfflieger
Oberleutnant Göring
654
Postkartenvertrieb W. Sanke
BERLIN N. 37.
Nachdruck wird gerichtlich verfolgt.

Unser erfolgreicher Kampfflieger
Oberleutnant Göring
655
Postkartenvertrieb W. Sanke
BERLIN N. 37
Nachdruck wird gerichtlich verfolgt.

Göttsch, Walter, *Leutnant*
Born Altour, 10 June 1896
Knight's Cross with Swords of the Royal Hohenzollern House Order, 23 August 1917
Sanke 518, 612, 613

FA 33, JASTA 8, 19
Killed in Action, 10 April 1918
20 Victories

612
Postkartenvertrieb W. Sanke
BERLIN N. 37
Nachdruck wird gerichtlich verfolgt.
Unser erfolgreicher Kampfflieger
Leutnant Göttsch

Unser erfolgreicher Kampfflieger
Leutnant Göttsch
613
Postkartenvertrieb W. Sanke
BERLIN N. 37.
Nachdruck wird gerichtlich verfolgt.

Greim, Robert Ritter von, *Oberleutnant*
Born Bayreuth, Bavaria, 22 June 1892
Pour le Mérite, 8 October 1918
Sanke 684

FFA 3, FAA 204, FA 46, JASTA 34, JGR 10, 9
Died by Suicide, 24 May 1945
28 Victories

Heibert, Robert, *Offizerstellvertreter*
Born Oberfall, 8 January 1886
Golden Military Merit Cross, 5 July 1918
Sanke 671, 672

FAA 207, 227, JASTA 33, 46
Died by Suicide, 10 May 1933
13 Victories

Unser erfolgreicher Kampfflieger
Offz.-Stellv. Heibert
672
Postkartenvertrieb W. Sanke
BERLIN N. 37.
Nachdruck wird gerichtlich verfolgt.

Heldman, Alois, *Leutnant*	FA 57, 59, JASTA 10
Born Gravenbruck, 2 December 1895	Died 1980s
Knight's Cross of the Royal Hohenzollern House Order, 1918	15 Victories
Sanke 664	

Hess, Ernst, *Leutnant*
Born Wiesbaden, 8 January 1893
Knight's Cross with Swords of the Royal Hohenzollern House Order, 5 December 1917
Sanke 590

FA 9 KEK DOUAI, JASTA 10, 28, 19
Killed in Action, 23 December 1917
17 Victories

Hesse, Hans, *Hauptmann*
Sanke 599

Hirth, Helmuth, *Leutnant*
Iron Cross Second Class
Sanke with no number, 422

Leutnant
Helmuth Hirth.
422
Postkartenvertrieb W. Sanke
BERLIN N. 37.
Nachdruck wird gerichtlich verfolgt.

Hoeppner, Ernst von, General du Kavalerie
Born Tonnin, 14 June 1860 Died, 25 September 1922
Pour le Mérite, 8 April 1917 Kommandierender General Army Air Service
Sanke 427, 505, 568, 600

Generalleutnant von Hoeppner
Kommandierender General der Luftstreitkräfte
505
Postkartenvertrieb W. Sanke
BERLIN N. 37.
Nachdruck wird gerichtlich verfolgt.

Generalleutnant von Hoeppner
Kommandierender General der Luftstreitkräfte
v. Hoeppner
568
Postkartenvertrieb W. Sanke
BERLIN N. 37
Nachdruck wird gerichtlich verfolgt.

Exzellenz von Hoeppner
kommandierender General der Luftstreitkräfte
und sein Generalstabs-Chef
Oberstlt. Thomsen
600
Postkartenvertrieb W. Sanke
BERLIN N. 37.
Nachdruck wird gerichtlich verfolgt.

Höhndorf, Walter, *Leutnant*
Born Prutzke, 10 November 1892
Pour le Mérite, 20 July 1916
Sanke 381, 389

FA 12, 67, KEK VAUX, JASTA 1, 4, 14
Killed in Flying Accident, 5 September 1917
12 Victories

Unser erfolgreicher Kampf-Flieger
Leutnant Höhndorf.
389
Postkartenvertrieb W. Sanke
BERLIN N. 37.
Nachdruck wird gerichtlich verfolgt.

Höhne, Otto Walter, *Leutnant*	KEK NORD, JASTA 1, 2, 59
Born Wiowitz, 30 April 1895	Died, 22 November 1969
Iron Cross First Class	6 Victories
Sanke 524, 525	

Leutnant Höhne.
525
Postkartenvertrieb W. Sanke
BERLIN N.37.
Nachdruck wird gerichtlich verfolgt.

Horn, Hans-George, *Oberleutnant*
Born Berbisdorf, 28 April 1892
Pour le Mérite, 23 December 1917
Sanke 601

Kagohl 221, 401

One of five observers awarded Pour le Mérite

Immelmann, Max, *Oberleutnant*
Born Dresden, 21 September 1890
Pour le Mérite, 12 January 1916
Sanke 340, 342, 347, 360, 361, 361 (Death Card), 362, 377, 610

FFA 62, KEK DOUAI
Killed in Action, 18 June 1916
15 Victories

Leutnant Immelmann
an den Trümmern eines von ihm abgeschossenen engl. Eindeckers.

Leutnant Immelmann
an seinem Fokker-Flugzeug
347
Postkartenvertrieb W. Sanke
BERLIN N. 37.
Nachdruck wird gerichtlich verfolgt

Leutnant Immelmann
360
Postkartenvertrieb W. Sanke
BERLIN N. 37
Nachdruck wird gerichtlich verfolgt.

Oberleutnant Immelmann
361
Postkartenvertrieb W. Sanke
BERLIN N. 37.
Nachdruck wird gerichtlich verfolgt.

Oberleutnant
Immelmann
361
Postkartenvertrieb W. Sanke
BERLIN N. 37.
Nachdruck wird gerichtlich verfolgt.

Leutnant Immelmann
362
Postkartenvertrieb W. Sanke
BERLIN N. 37

Immelmann
Stark und sicher als deutscher Aar
Führt'st Du Dein Flugzeug durch Not und Gefahr,
Schwebtest in Lüften ruhig und leicht,
Wolkenbezwinger, von niemand erreicht!
An Jahren so jung, an Mut so groß
Blühender Lorbeer in Deinem Schooß!
Immelmann, kühnster der Flieger!
Immelmann, stolzester Sieger!
Wie Dein leuchtendes Flügelpaar
Furcht und Schrecken der Feinde war
Wie Du — entsetzlichster Kampf in den Lüften
Aus schwindelnder Höhe zu tiefsten Grüften
Todtrotzend die Gegner bezwungen
Sei später begeistert gesungen!
Immelmann, kühnster der Flieger
Immelmann, stolzester Sieger!
Heut lastet die Trauer zu groß und schwer
Unser Stolz, unser Adler, er fliegt nicht mehr!
Sein Flug zur leuchtenden Sonne hinauf
Brach jäh ihm das Herz im Siegeslauf!
Das Ehrenzeichen, der Lorbeer grün,
Wird stolz auf dem Grabe des Helden blühn,
Dankend und trauernd mit liebender Hand
Hütet und pflegt ihn das Vaterland.
Immelmann, kühnster der Streiter
Lebt leuchtend im Volke weiter.
377
Postkartenvertrieb W. Sanke
BERLIN N 37

Oberleutnant Immelmann †
(letzte Aufnahme)
610
Postkartenvertrieb W. Sanke
BERLIN N 37
Nachdruck wird gerichtlich verfolgt.

Jacobs, Josef Carl Peter, *Leutnant*
Born Kreuzkapelle, 15 May 1894
Pour le Mérite, 18 July 1918
Sanke 618, 622, 680

FFA 11, JASTA 12, 22, 7
Died, 29 July 1978
48 Victories

Unser erfolgreicher Kampfflieger
Leutnant Jacobs
622
Postkartenvertrieb W. Sanke
BERLIN N 37
Nachdruck wird gerichtlich verfolgt.

Unser erfolgreicher Kampfflieger
Leutnant Jacobs
680
Postkartenvertrieb W. Sanke
BERLIN N. 37.
Nachdruck wird gerichtlich verfolgt.

Jörke, Reinhold, *Offizierstellvertreter* JASTA 12, 13, 39
Iron Cross First Class 14 Victories
Sanke 591

Keller, Alfred, *Hauptmann*
Born Bochun, 19 September 1882
Pour le Mérite, 4 December 1917
Sanke 602

KAGOHL 27, AFP 5, KAGOHL 40, BAGOHL 1
Died, 11 February 1974
Commanded BAGOHL 1

Keudell, Hans von, *Leutnant*
Born Berlin, 5 April 1892
Knight's Cross with Swords of the Royal Hohenzollern House Order, 1 January 1917
Sanke 555, 556

BAO, KEK B, JASTA 1, 27
Killed in Action, 15 February 1917
12 Victories

Leutnant Parschau u. Leutnant von Keudell

Kirmaier, Stefan, *Oberleutnant*
Born Lachen, 28 July 1889
Knight's Cross with Swords of the Royal Hohenzollern House Order, 22 November 1916
Sanke 445

FAA 203, KEK J, JASTA 2
Killed in Action, 22 November 1916
11 Victories

Kirschstein, Hans, *Leutnant*
Born Koblenz, 5 August 1896
Pour le Mérite, 24 June 1918
Sanke 623

FA 19, 256, 3, JASTA 6
Accidentally killed, 16 July 1918
27 Victories

Kissenberth, Otto, *Oberleutnant*
Born Landshut, 26 February 1893
Pour le Mérite, 24 July 1918
Sanke 595

FA 8, 9, KEK E, JASTA 16, 23
Killed Mountaineering, 2 August 1919
20 Victories

Klein, Hans, *Oberleutnant*
Born Stettin, 17 January 1891
Pour le Mérite, 2 December 1917
Sanke 514, 515, 592, 593

JASTA 4, 10
Died, 18 November 1944, Luftwaffe Major General
22 Victories

Unser erfolgreicher Kampfflieger
Leutnant Klein.
515
Postkartenvertrieb W. Sanke
BERLIN N. 37.
Nachdruck wird gerichtlich verfolgt

Unser erfolgreicher Ka pfflie ger
Leutnant Klein
592
Postkartenverlag W. Sanke
BERLIN N 37
Nicola Perscheid phot.

Unser erfolgreicher Kampfflieger
Leutnant Klein
593
Postkartenvertrieb W. Sanke
BERLIN N. 37
Nachdruck wird gerichtlich verfolgt.
Nicola Perscheid phot.

Kleine, Rudolf, Hauptmann	KAGOHL 1, 3
Born Minden, 28 August 1886	Killed in Action, 12 December 1917
Pour le Mérite, 4 October 1917	Led bombing raids on London
Sanke 545	

Klimke, Rudolf, *Leutnant* FFA 55, FA 50, 19, KASTA 13, JASTA 27
Born Merseberg, 8 November 1890
Iron Cross First Class 17 Victories
Sanke 674

Könnecke, Otto, *Leutnant* — JASTA 25, 5
Born Strasbourg, 20 September 1892 — Died, 25 January 1956
Pour le Mérite, 26 September 1918 — 35 Victories
Sanke 643

Kosmahl, Fritz, *Offizierstellvertreter*
Born Leipzig, 5 September 1892
Prussian Member's Cross with Swords
of the Royal Hohenzollern House Order, 11 January 1915
Sanke 546

FAA 22, 261, JASTA 26
Died of wounds, 26 September 1917
9 Victories

Kroll, Heinrich Cladius, *Oberleutnant* FA 17, JASTA 9, 24
Born Flatsby, Flensberg, 3 November 1894 Died, 21 February 1930
Pour le Mérite, 29 March 1918 33 Victories
Sanke 597, 625, 626

Unser erfolgreicher Kampfflieger
Leutnant Kroll
625
Postkartenvertrieb W. Sanke
BERLIN N. 37
Nachdruck wird gerichtlich verfolgt.

Unser erfolgreicher
Leutnant Kroll
626
Postkartenvertrieb W. Sanke
BERLIN N. 37

Kypke, Walter, *Leutnant*
Born Stettin, 22 September 1893
Knight's Cross of the Royal Hohenzollern House Order
Sanke 678

FFA 2, 58, KEK A, JASTA 14, 41, KEST 5, JASTA 47
Died in air crash, 6 June 1924
9 Victories

Laumann, Arthur, *Leutnant*
Born Essen, 4 July 1894
Pour le Mérite, 25 October 1918
Sanke 668, 670

FAA 265, JASTA 66, 10
Died, 18 November 1970
28 Victories

Unser erfolgreicher Kampfflieger
Leutnant Laumann
670
Postkartenvertrieb W. Sanke
BERLIN N. 37
Nachdruck wird gerichtlich verfolgt.

Leffers, Gustav, *Leutnant*
Born Wilhelmshaven, 2 January 1892
Pour le Mérite, 5 November 1916
Sanke 372, 413, 414, 415, 417

FFA 32, KEK B, JASTA 1
Killed in Action, 27 December 1916
9 Victories

Unser erfolgreicher Kampf-Flieger
Leutnant Leffers
413
Postkartenvertrieb W. Sanke
BERLIN N. 37.
Nachdruck wird gerichtlich verfolgt.

Unser erfolgreicher Kampf-Flieger
Leutnant Leffers
414
Postkartenvertrieb W. Sanke
BERLIN N. 37.
Nachdruck wird gerichtlich verfolgt.

Unser erfolgreicher Kampf-Flieger
Leutnant Leffers
415
Postkartenvertrieb W. Sanke
BERLIN N. 37
Nachdruck wird gerichtlich verfolgt

Unser erfolgreicher Kampf-Flieger
Leutnant Leffers
417
Postkartenvertrieb W. Sanke
BERLIN N. 37.
Nachdruck wird gerichtlich verfolgt.

Loerzer, Bruno, *Hauptmann*
Born Berlin, 22 January 1891
Pour le Mérite, 12 March 1917
Sanke 616, 651

FAA 25, FA 60, FAA 203, KEK J, JASTA 17, 26, JG III
Died, 23 August 1960
44 Victories

651
Postkartenvertrieb W. Sanke
BERLIN N. 37.
Nachdruck wird gerichtlich verfolgt.
Unser erfolgreicher Kampfflieger
Oberleutnant Loerzer

Löwenhardt, Erich, *Oberleutnant*
Born Breslau, 7 April 1897
Pour le Mérite, 31 May 1918
Sanke 614, 641, 642

FAA 265, JASTA 10
Killed in Collision, 10 August 1918
54 Victories

Unser erfolgreicher Kampfflieger
Leutnant Loewenhardt
641
Postkartenvertrieb W. Sanke
BERLIN N. 37.
Nachdruck wird gerichtlich verfolgt.

Unser erfolgreicher Kampfflieger
Leutnant Loewenhardt
642
Postkartenvertrieb W. Sanke
BERLIN N. 37.
Nachdruck wird gerichtlich verfolgt.

Mai, Josef, *Leutnant* KASTA 29, JASTA 5
Born Berlin, 3 March 1887 Died, 18 January 1982
Golden Merit Military Cross, 24 June 1918 30 Victories
Sanke 660

Manschott, Friedrich, *Vizefeldwebel*
Born Reichartshausen, 21 February 1893
Iron Cross First Cross
Sanke 448, 449

FAA 203, JASTA 7
Killed in Action, 16 March 1917
12 Victories

Unser erfolgreicher Kampf-Flieger
Vfw. Manschott
449
Postkartenvertrieb W. Sanke
BERLIN N. 37.
Nachdruck wird gerichtlich verfolgt

Menckhoff, Carl, *Oberleutnant* JASTA 3, 72
Born Herford, 14 April 1883 Died 1948
Pour le Mérite, 23 April 1918 39 Victories
Sanke 617, 634

Unser erfolgreicher Kampfflieger
Leutnant Menckhoff
634
Postkartenvertrieb W. Sanke
BERLIN N. 37
Nachdruck wird gerichtlich verfolgt.
Nicola Perscheid
phot.

Meyer, Karl, *Oberflugmeister*
Born Mulhausen, 29 January 1894
Iron Cross First Class
Sanke 411

SFS 1, MFJ 1
Accidentally killed, 31 December 1917
8 Victories

Müller, Hans, *Leutnant* JASTA 12, 15, 18
Born Etzell, 3 July 1896
Iron Cross First Class 12 Victories
Sanke 446, 447

Unser erfolgreicher Kampf-Flieger
Leutnant Hans Müller
447
Postkartenvertrieb W. Sanke
BERLIN N. 37
Nachdruck wird gerichtlich verfolgt.

Müller, Max Ritter von, *Leutnant*
Born Rottenburg, 1 January 1887
Pour le Mérite, 3 November 1917
Sanke 444, 552, 553, 583, 584

FA1, KEK B, JASTA 2, 28
Killed in Action, 9 January 1918
36 Victories

Unser erfolgreicher Kampfflieger
Leutnant Max Müller
552
Postkartenvertrieb W. Sanke
BERLIN N 37

Unser erfolgreicher Kampfflieger
Leutnant Max Müller
553
Postkartenvertrieb W.Sanke
Nachdruck wird

Unser erfolgreicher Kampfflieger
Leutnant Max Müller
583
Postkartenvertrieb W. Sanke
BERLIN N. 37
Nachdruck wird gerichtlich verfolgt.

Unser erfolgreicher Kampfflieger
Leutnant Max Müller
584
Postkartenvertrieb
BERLIN
Nachdruck wird

Mulzer, Max Ritter von, *Leutnant*
Born Bavaria, 9 July 1893
Pour le Mérite, July 1916
Sanke 379, 385, 396, 397, 398

FFA 4, FA 62, KEK N, KEK B, FA 32
Accidentally killed, 26 September 1916
10 Victories

Unser erfolgreicher Kampf-Flieger
Leutnant Mulzer.
385
Postkartenvertrieb W. Sanke
BERLIN N. 37.
Nachdruck wird gerichtlich verfolgt.

Unser erfolgreicher Kampf-Flieger
Leutnant Mulzer
396
Postkartenvertrieb W. Sanke
BERLIN N. 37.
chdruck wird gerichtlich verfolgt.

Unser siegreicher Kampf-Flieger
Leutnant Mulzer
397
W. Sanke
BERLIN N.37

Unser erfolgreicher Kampf-Flieger
Leutnant Mulzer
398
Postkartenvertrieb W. Sanke
BERLIN N. 37.
Nachdruck wird gerichtlich verfolgt.

Nathanael, Edmund, *Offizierstellverteter*
Born Dielsdorf, 18 December 1889
Wilhelm Ernst War Cross, 1916
Sanke 537

FFA 42, JASTA 22, 5
Killed in Action, 11 May 1917
15 Victories

Neckel, Ulrich, *Leutnant*
Born, 23 January 1898
Pour le Mérite, 8 November 1918
Sanke 678

FA 25, JASTA 12, 19, 6
Died, 11 May 1928
30 Victories

Nerger, *Fregattenkapitan*
Pour le Mérite
Sanke 632, 633

Commanded the auxillary cruiser and raider "Wolf"

Fregattenkapitän Nerger
633
Postkartenvertrieb W. Sanke
BERLIN N. 37.
Nachdruck wird gerichtlich verfolgt.
Nicola Perscheid
phot.

Nielebock, Friedrich, *Leutnant*	FA 250
Born Weissenwarthe, 4 July 1882	
Pour le Mérite, 2 June 1918	280 Sorties as an Observer
Sanke 649	

Parschau, Otto, *Leutnant*	FFA 42, 261, KG 1, FA 32
Born Klutznitz, 11 November 1890	Died of Wounds, 21 July 1916
Pour le Mérite, 10 July 1916	8 Victories
Sanke 380	

Pechmann, Paul Freiherr von, *Oberleutnant*
Born Gaudismühl, 28 December 1889
Pour le Mérite, 31 July 1917
Sanke 551

FFA 6, FA 215, 217

700 Sorties as an Observer

Pfeiffer, Hermann, *Leutnant*
Born Tengen, 24 July 1890
Militarischer Karl-Friedrich Verdeinst-Orden, 21 November 1916
Sanke 428, 451

FFA 10, AOK 3, JASTA 9
Accidentally killed, 20 May 1917
11 Victories

Unser erfolgreicher Kampf-Flieger
Leutnant Pfeifer
451
Postkartenvertrieb W. Sanke
BERLIN N. 37.
Nachdruck wird gerichtlich verfolgt.

Pippart, Hans Martin, *Leutnant*
Born Mannheim, 14 May 1888
Knight's Cross with Swords of the Royal Hohenzollern House Order, 2 May 1918
Sanke 624

FAA 220, KASTA 1, JASTA 13, 19
Killed in Action, 11 August 1918
22 Victories

Plüschow, Günther, *Kapitanleutnant*
Born Munich, 8 February 1886
Knight's Cross with Swords of the Royal Hohenzollern House Order
Sanke 550

Served in China, 1914
Died, 28 January 1931

Pütter, Fritz, *Leutnant*
Born Duelmen, 14 January 1895
Pour le Mérite, 31 May 1918
Sanke 627

FA 251, JASTA 9, 68
Died of Injuries, 10 August 1918
25 Victories

Ray, Franz, *Leutnant* JASTA 1, 28, 49
Staffelführer JASTA 49, 15 December1917
17 Victories
Sanke 658

Reimann, Leopold Rudolf, *Offizierstellvertreter*
Born Oberhohnsdorf, 7 December 1892
Iron Cross First Class, 10 September 1916
Sanke 429

KEK B, JASTA 1, 2
Accidentally killed, 24 January 1917
5 Victories

Richthofen, Lothar Freiherr von, *Oberleutnant*
Born Breslau, 27 September 1894
Pour le Mérite, 14 May 1917
Sanke 526, 656

KG 4, JASTA 11
Died in a flying accident, 4 July 1922
40 victories

656
Postkartenvertrieb W. Sanke
BERLIN N. 37
Nachdruck wird gerichtlich verfolgt.
Lothar Freiherr von Richthofen

Richthofen, Manfred Freiherr von, *Rittmeister*
Born Breslau, 2 May 1892
Pour le Mérite, 12 January 1917
Sanke 450, 503, 509, 519, 532, 533, 533, 534, 534, 554, 619

FA 69, KG 2, JASTA 2, 11, JG 1
Killed in Action, 21 April 1918
80 Victories

Rittmeister Manfred Frhr. von Richthofen
Phot. C. J. von Dühren.
BERLIN
503
Postkartenvertrieb W. Sanke
BERLIN N. 37.
Nachdruck wird gerichtlich verfolgt.

Der kommandierende General der Luftstreitkräfte
Exz. von Hoeppner
beglückwünscht Rittmeister Frhr. von Richthofen.
509
Postkartenvertrieb W. Sanke
BERLIN N. 37

Oberstleutnant Thomsen
Chef des Generalstabes
der Luftstreitkräfte
Rittmeister
Freiherr v. Richthofen
unser erfolgreichster Flieger
Generalleutnant von Hoeppner
kommandierender General
der Luftstreitkräfte
519
Postkartenvertrieb W. Sanke
„BERLIN N. 37„
Nachdruck wird gerichtlich verfolgt.

Rittmeister Manfred Frhr. von Richthofen
532
Postkartenvertrieb W. Sanke
BERLIN N. 37.
Phot. Nicola Perscheid
Nachdruck wird gerichtlich verfolgt.
1918

Rittmeister Manfred Frhr. von Richthofen
533
Postkartenvertrieb W.Sanke
BERLIN N.37.
Nachdruck wird gerichtlich verfolgt

Rittmeister
Manfred Frhr. von Richthofen ✝
Nicola Perscheid
phot.
533
Postkartenvertrieb W. Sanke
BERLIN N. 37
Nachdrück wird gerichtlich verfolgt.

Rittmeister Manfred Frhr. von Richthofen
534
Postkartenvertrieb W. Sanke
BERLIN N. 37
Nachdruck wird gerichtlich verfolgt

Rittmeister
Manfred Frhr. von Richthofen +
Nicola Perscheid
phot.
534
Postkartenvertrieb W. Sanke
BERLIN N. 37
Nachdruck wird gerichtlich verfolgt.

Unsere beiden erfolgreichsten Flieger
Rittmeister Frhr. von Richthofen
u. Leutnant Voss

Rittmeister Manfred Frhr. von Richthofen †

Richthofen, Richthofen, König der Luft!
Richthofen, höre, Dein Kaiser ruft! — —
Schon steigt er zur Höhe!
Schon siegt er im Licht!
Das Rufen des Kaisers selbst hört er dort nicht!
Hoch oben im Lichtmeer
Dort ist sein Reich!
Dort herrscht er als König und Kaiser zugleich!

So sehn wir den Helden, so bleibt er uns nah,
Ob auch das Furchtbare jetzt geschah!
Ein Siegen und Kämpfen in schwindelnden Höhn
Ein Jagen durch Lüfte wie wirbelnder Föhn!

Der achtzigste Sieg! — — Und das ganze Land
Reicht sich in Trauer still die Hand. — —
Richthofen, Richthofen, Dein Glorienschein
Wird unter großen der größeste sein!

Ruth. Barbara Goedr.

619
Postkartenvertrieb W. Sanke
BERLIN N 37

Reiper, Peter, *Leutnant*	Ballonzug 19
Born Hannover, 13 April 1887	
Pour le Mérite, 7 July 1918	Only Pour le Merìte Balloon Observer
Sanke 659	

Rosencrantz, Hans, *Leutnant*	KG 1
Born Wollstein, 9 August 1890	Killed in Action, 6 September 1916
Iron Cross First Class	5 Victories as an Observer
Sanke 419	

Röth, Friedrich Ritter von, *Oberleutnant*
Born Nürnberg, 29 September 1893
Pour le Mérite, 9 September 1918
Sanke 647

FAA 296, JASTA 34, 23, 16
Committed Suicide, 31 December 1918
28 Victories

Rumey, Fritz, *Leutnant*
Born Königsberg, 3 March 1891
Pour le Mérite, 10 July 1918
Sanke 639, 665, 666, 667

FAA 19, JASTA 2, 3
Killed in Action, 27 September 1918
45 Victories

665
Postkartenvertrieb W. Sanke
BERLIN N 37
Nachdruck wird gerichtlich verfolgt
Unser erfolgreicher Kampfflieger
Leutnant Rumey

Unser erfolgreicher Kampfflieger
Leutnant Rumey
666
Postkartenvertrieb W. Sanke
BERLIN N. 37
Nachdruck wird gerichtlich verfolgt.

Unser erfolgreicher Kampfflieger
Leutnant Rumey
667
Postkartenvertrieb W. Sanke
BERLIN N 37.

Schäfer, Karl Emil, *Leutnant*
Born Krefeld, 17 December 1891
Pour le Mérite, 26 April 1917
Sanke 507, 507, 512

KASTA 8, 11 JASTA 11, 28
Killed in Action, 5 June 1917
30 Victories

507

Leutnant Schaefer

512
Postkartenvertrieb W. Sanke
BERLIN N. 37
Nachdruck wird gerichtlich verfolgt
Leutnant Schäfer.

Schleiffer, Josef *Vizefeldwebel* SCHUSTA 7
Pilot for Hauptmann Zorer
Iron Cross First Class
Sanke 538

Schleich, Eduard Ritter von, *Hauptmann*
Born Munich, 9 August 1888
Pour le Mérite, 4 December 1917
Sanke 578, 596, 644, 646

FA 2, SS 28, JASTA 21, 32, JGr 8
Died, 15 November 1947
35 Victories

Oberleutnant Schleich
596
Postkartenvertrieb W. Sanke
BERLIN N. 37.
Nachdruck wird gerichtlich verfolgt

Hauptmann Ritter von Schleich
644
Postkartenvertrieb W. Sanke
BERLIN N.

Hauptmann Ritter von Schleich
im Gespräch mit Leutnant Röth
646
BERLIN N 37

Schmidt, Julius, *Leutnant*
Place and date of birth, unknown
Knight's Cross with Swords of the Royal Hohenzollern House Order, November 1917
Sanke 575

KG 4, JASTA 3, 6
Died, 2 July 1944
15 Confirmed Victories

Schmidt, Otto, *Oberleutnant*
Born Neukirchen, 23 March 1885
Knight's Cross with Swords of the Royal Hohenzollern House Order, 19 September 1917
Sanke 628

FA 25, JASTA 7, 32, 29, 5, JGr 11, 2

20 Victories

Schreiber, Wilhelm Paul, *Leutnant*
Knight's Cross with Swords of the Royal Hohenzollern House Order, 7 March 1918
Pour le Mérite Posthumously, 8 June 1918
Sanke 661

FAA 221
Killed in Action, 30 May 1918

Schulte, Adolf, Leutnant
Born Ludenscheit, 24 December 1894
Iron Cross First Class
Sanke 547

JASTA 12
Killed in Action, 12 April 1917
9 Victories

Unser erfolgreicher Flieger
Rittmstr. Prinz Friedrich Sigismund v. Preußen

Sigismund, Prinz Friedrich, *Rittmeister*
Son of Kaiser Wilhelm II
Iron Cross First Class
Higher Awards by Birth
Sanke 569, 570, 579, 580, 581

Unser erfolgreicher Flieger
Rittmstr. Prinz Friedrich
Sigismund v. Preußen
570
Postkartenvertrieb W. Sanke
BERLIN N. 37.
Nachdruck wird gerichtlich verfolgt.

Unser erfolgreicher Flieger
Rittmstr. Prinz Friedrich
Sigismund v. Preußen
Photo: Karl Selberg.
579
Postkartenvertrieb W. Sanke
BERLIN
Nachdruck wird gerichtlich verfolgt.

Unser erfolgreicher Flieger
Rittmstr. Prinz Friedrich
Sigismund v. Preußen
Photo: Karl Selberg.
580
Postkartenvertrieb W. Sanke
BERLIN N. 37.
Nachdruck wird gerichtlich verfolgt.

Unser erfolgreicher Flieger
Rittmstr: Prinz Friedrich
Sigismund v. Preußen
581
Postkartenvertrieb W. Sanke
BERLIN N 37.
Nachdruck wird gerichtlich verfolgt

Stabbert, Franz, *Kapitänleutnant*

Sanke 574

Under Stabbert's command L 44 was destroyed over France
Killed in Action, 20 October 1917
Captained L7, 1915; L20, 1915; L23, 1916; L44, 1917

Stein, Alexander, *Leutnant* Naval Observation

With *Oberflugmeister* Paul Fabeck, flew from the ocean raider “Wolf”

Iron Cross First Class, 1917

Sanke 637, 640

Die beiden Wolfsflieger
Leutnant d. R. Alexander Stein
Oberflugmeister d. S. I. Paul Fabeck
640
Postkartenvertrieb W. Sanke
BERLIN N. 37.
Nachdruck wird gerichtlich verfolgt.

Strasser, Georg, *Vizefeldwebel* FA 44, JASTA 17
Died, 4 December 1925
Iron Cross First Class 7 Victories
Sanke 638

Strasser, Peter, *Fregattenkapitän*
Born Hannover, 1 April 1876
Pour le Mérite, 4 September 1917
Sanke 567

Chief, Naval Airship Service
Killed in Action, 5 August 1918
Raids Over England

Thom, Karl, *Leutnant*
Born Freystadt, 19 May 1893
Pour le Mérite, 1 November 1918
Sanke 645

FFA 216, 48, 234, JASTA 21
Missing in Action, 1943
27 Victories

Thomson, Hermann, *Oberstleutnant* Chief of Staff of the Commanding Generals of the Luftstreitkräfte
Pour le Mérite
Sanke 504

Tutschek, Adolf Ritter von, *Hauptmann*
Born Ingolstadt, 16 May 1891
Pour le Mérite, 3 August 1917
Sanke 540, 572, 650

FA 6, JASTA 2, 12, J. G. II
Killed in Action, 15 March 1918
27 Victories

Unser erfolgreicher Kampfflieger
Oberleutnant Ritter von Tutschek
572
Postkartenvertrieb W. Sanke
BERLIN N. 37.
Nachdruck wird gerichtlich verfolgt.

650
Postkartenvertrieb W. Sanke
BERLIN N 37
Nachdruck wird gerichtlich verfolgt
Hauptmann Ritter von Tutschek
Kommandeur eines Jagdgeschwaders

Thuy, Emil, *Leutnant*
Born Hagen, 11 March 1894
Pour le Mérite, 30 June 1918
Sanke 577

FFA 53, JASTA 21, 28, JGr 7
Killed in an air crash, 11 June 1930
35 Victories

Udet, Ernst, *Oberleutnant*
Born Frankfurt-am-Main, 26 April 1896
Pour le Mérite, 23 April 1918
Sanke 604, 620, 621, 652, 653

FAA 206, FA 68, KEK H, JASTA 15, 37, 11, 4
Committed Suicide, 17 November 1941
62 Victories

620
Postkartenvertrieb W. Sanke
BERLIN N. 37.
Nachdruck wird gerichtlich

Unser erfolgreicher Kampfflieger
Leutnant Udet
621
Postkartenvertrieb W. Sanke
BERLIN N. 37
Nachdruck wird gerichtlich verfolgt
phot. A. Buck, München

Unser erfolgreicher Kampfflieger
Leutnant Udet
652
Postkartenvertrieb W. Sanke
BERLIN N. 37.
Nachdruck wird gerichtlich verfolgt.

653
Postkartenvertrieb W. Sanke
BERLIN N. 37.
Nachdruck wird gerichtlich verfolgt.
Unser erfolgreicher Kampfflieger
Leutnant Udet

Ulbrick, *Flugobermaat*
Iron Cross First Class
Sanke 544

Ultsch, Bernhard, *Offizierstellvertreter*
Born Wunseidel, 26 March 1898
Bavarian Military Merit Cross Second Class
with Swords, September 1918
Sanke 663

SCHUSTA 29, JASTA 39, 77

12 Victories

Veltjens, Joseph, *Leutnant*
Born Geldern, 2 June 1894
Pour le Mérite, 16 August 1918
Sanke 681

FA 23, JASTA 14, 18, 15
Killed in Action, 6 October 1943
35 Victories

Voss, Werner, *Leutnant*
Born Krefeld, 13 April 1897
Pour le Mérite, 8 April 1917
Sanke 506, 520, 554

KG 4, JASTA 2, 5, 29, 14, 10
Killed in Action, 23 September 1917
48 Victories

Unser erfolgreicher Kampfflieger
Leutnant Voß.

Unsere beiden erfolgreichsten Flieger
Rittmeister Frhr. von Richthofen
u. Leutnant Voss
554
Postkartenvertrieb W. Sanke
BERLIN N. 37

Wendelmuth, Rudolf, *Leutnant*
Born Gotha, 28 July 1890
Iron Cross First Class
Sanke 594

FA 5, JASTA 8, 20
Died in Collision, 30 November 1917
14 Victories

Leutnant Windisch

Windisch, Rudolf, *Leutnant*
Born Dresden, 27 January 1897
Pour le Mérite, 6 June 1918
Sanke 576, 662

FAA 62, KG 2, JASTA 32, 50, 66
Missing in Action, 27 May 1918
22 Victories

Leutnant Windisch

Wintzens, Kurt, *Leutnant*
Born Neustadt, 1 August 1894
Pour le Mérite, 1 July 1916
Sanke 378, 391, 392, 400

FFA 67, 6, 48, KEK VAUX, JASTA 4, 1
Killed in Action, 25 September 1916
19 Victories

Unser erfolgreicher Kampf-Flieger
Leutnant Wintgens
391
Postkartenvertrieb W. Sanke
BERLIN N. 37.
Nachdruck wird gerichtlich verfolgt.

Unser erfolgreicher Kampf-Flieger
Leutnant Wintgens
392
Postkartenvertrieb W. Sanke
BERLIN N. 37
Nachdruck wird gerichtlich verfolgt.

Direktor Fokker und
Leutnant Wintgens.

Wolff, Kurt, *Oberleutnant*
Born Greifswold, 6 February 1895
Pour le Mérite, 4 May 1917
Sanke 513, 522, 523, 542

JASTA 11, 29
Killed in Action, 15 September 1917
33 Victories

Leutnant Wolff.
522
Postkartenvertrieb W. Sanke
BERLIN N. 37
Nachdruck wird gerichtlich verfolgt.

Leutnant Wolff.
523
Postkartenvertrieb W. Sanke
BERLIN N. 37.
Nachdruck wird gerichtlich verfolgt.

A-333
A-333
Leutnant Wolff
542
Postkartenvertrieb W. Sanke
BERLIN N. 37.
Nachdruck wird gerichtlich verfolgt.

Wusthoff, Kurt, *Leutnant*
Born Aachen, 27 January 1897
Pour le Mérite, 26 November 1917
Sanke 560, 563, 564, 566, 586, 587, 588

KG 2, JASTA 4, 15
Killed in Air Show Accident, 23 July 1926
27 Victories

Unser erfolgreicher
Kampf-Flieger
Leutnant Wüsthoff
563
Postkartenvertrieb W.
BERLIN N 37
Nachdruck wird gerichtlich verfolgt.

Unser erfolgreicher Kampfflieger
Leutnant Wüsthoff
564
Postkartenvertrieb W. Sanke
BERLIN N. 37.
Nachdruck wird gerichtlich verfolgt.

Unser erfolgreicher Kampfflieger
Leutnant Wüsthoff
566
Postkartenvertrieb W. Sanke
BERLIN N. 37.
Nachdruck wird gerichtlich verfolgt.

Unser erfolgreicher Kampfflieger
Leutnant Wüsthoff
586
Postkartenvertrieb W. Sanke
BERLIN N. 37.
Nachdruck wird gerichtlich verfolgt.
Phot. Hugo Erfurth, Dresden.

Unser erfolgreicher Kampfflieger
Leutnant Wüsthoff
587
Postkartenvertrieb W. Sanke
BERLIN N. 37
Nachdruck wird gerichtlich verfolgt
Phot. Hugo Erfurth, Dresden.

Unser erfolgreicher Kampfflieger
Leutnant Wüsthoff
588
Postkartenvertrieb W. Sanke
BERLIN N 37
Nachdruck wird gerichtlich verfolgt
Phot. Hugo Erfurth, Dresden

Zander, Martin, *Hauptmann* FAA 90, KEK NORD, JASTA 1
Died, 1925
Iron Cross First Class 5 Victories
Sanke 407

Zorer, Eduard Wolfgang, *Hauptmann*
Knight's Cross with Swords of the Royal Hohenzollern House Order, 5 August 1917
Sanke 539

FAA 54, Commanded SCHUSTA 7

Early Iron Second Class Cross Winners FFA 26
Sanke 313

Hauptmann Boelcke und *Oberleutnant* Immelmann (Group of 13 Airmen)
Sanke 373

Die fünf erfolgreichsten Flieger der Jagdstaffel Richthofen ***Sanke 371*** *and* ***511***

Unsere Flieger-Helden ***Sanke 408***

Jagdstaffel Richthofen ***Sanke 606***

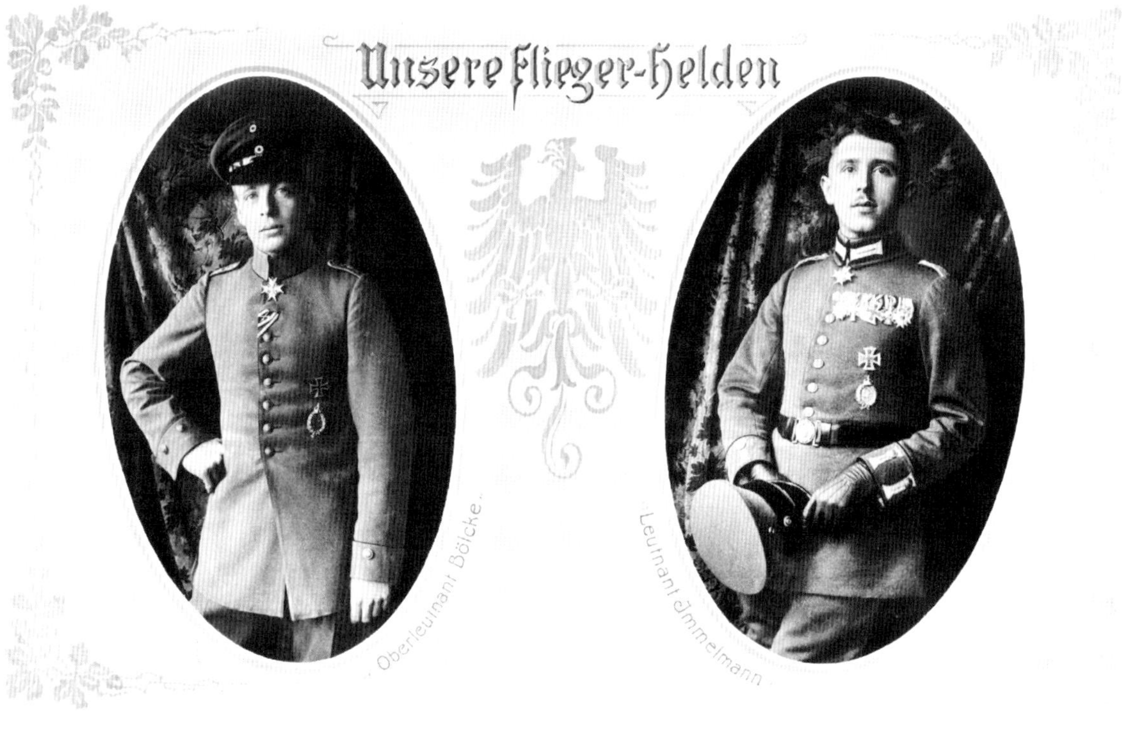

Boelcke and Immelmann ***Sanke 364***

Bibliography

Aviation Awards of Imperial Germany, seven volumes by Neal W. O'Connor

Volume I: *The Aviation Awards of the Kingdom of Bavaria*. Published by the Foundation for Aviation World War I. Princeton, New Jersey, 1988.

Volume II: *The Aviation Awards of the Kingdom of Prussia*. Published by the Foundation of Aviation World War I. Princeton, New Jersey, 1990. pp 285. Profusely illustrated.

Volume III: *The Aviation Awards of the Kingdom of Saxony*. The Foundation for Aviation World War I. Princeton, New Jersey. 1993. pp 279. Profusely illustrated.

Volume IV: *The Aviation Awards of the Kingdom of Württemberg*. Foundation for Aviation World War I. Princeton, New Jersey; and the Flying Machine Press. Mountain View, California. 1995. pp 280. Illustrations and aircraft profiles.

Volume V: *The Aviation Awards of the Eight Thuringian States and the Duchy of Anholt*. Foundation for Aviation World War I. Princeton, New Jersey, and the Flying Machine Press, Stratford, Connecticut. 1998. pp 364. Profuse illustrations and color aircraft profiles.

Volume VI: *The Aviation Awards of the Grand Duchies of Baden and Oldenburg*. Foundation for Aviation World War I, Princeton, New Jersey; and the Flying Machines Press, Stratford, Connecticut. 1999. pp 428. Profusely illustrated.

Volume VII: *The Aviation Awards of Eight German States and the three Free Cities*. Schiffer Military History. Atglen, Pennsylvania. 2002. pp 400. Over 700 photographs.

Neal W. O'Connor's epic seven volumes tell in detail of the lives, victories, and awards of Germany's most prominent aviators. The illustrations, both formal and from the field, are unsurpassed and many of Sanke's cards are used throughout. Neal O'Connor sadly passed away in 2002, but his work will remain the mother lode on the subject.

Above the Lines, A Complete Record of the Fighter Aces of the German Air Service, Naval Air Service and Flanders Marine Corps, 1914-1918. Norman L. R. Franks, Frank W. Bailey, Russell Guest. Grub Street, London. 1993. pp 259. Many illustrations, including Sanke images.

Casualties of the German Air Service 1914-1920. Norman Franks, Frank Bailey, Rick Duiven. Grub Street. London. 1999. pp. 384. No illustrations.

Fliegertruppe 1914-1918. Published by the author, A. E. Ferko, Salem, Ohio. 1980. pp 48. Illustrated by photographs from the author's extensive collection now at the Aviation History Collection, University of Texas, Dallas.

Fliegertruppe 1914-1918, Nr. 2. Published by the author, A. E. Ferko. Salem, Ohio. 1986. pp 48. With numerous photos from the Ferko collection.

The Jasta Pilots, Detailed listings and Histories, August 1916-November 1918. Norman Franks, Frank Bailey, Rick Duiven. Grub Street. London. 1996. pp 364; pp 32 of illustrations, some by Sanke.

The Red Baron Combat Wing, Jagdgeschwader Richthofen in Battle. Peter Kilduff. Arms and Armor Press. London. 1997. pp 288. Many photos, daily victory and casualty lists of JG 1 units and an extensive bibliography.

Knight of Germany Oswald Boelcke – German Ace. Professor Johannes Werner. John Hamilton Ltd. London. 1933. pp 241. 27 illustrations and 4 maps.

Cross and Cockade Journal. The Society of World War Aero Historians. Various issues.

Cross and Cockade – Great Britain. Various issues.

Over The Front. Journal of the League of World War I Aviation Historians. Various issues.

Richthofen, Beyond the Legend of the Red Baron. Peter Kilduff. John Wiley and Sons. New York. 1993. pp 256; pp 32 of Illustrations.

The Illustrated Red Baron, The Life and Times of Manfred von Richthofen. Peter Kilduff. Cossell and Company. London. 1999. pp 160. Profusely illustrated. Pilot rosters for all the units of Jagdgeschwader Richthofen.

Flying With Richthofen (Jagd in FlandersHimmel) Karl Bodenschatz. Translated by Jan Hayzlett. Grub Street. London. 1996. pp 224; pp 16 of Illustrations, War Journal of Jagdgeschwader Nr. 1 June 1917 – 11 November 1918.

The Jasta War Chronology. A Complete Listing of Claims and Losses, August 1916-November 1918. Norman Franks, Frank Bailey, Rick Duiven. Grub Street. London. 1998. pp 285, pp 24, illustrations.

German Language

Flieger an allen Fronten. Friedrich Schilling. Verlag Scherl. Berlin. 1936. pp 224, 44 illustrations.

Immelmann der Adler von Lifle. Edited by his brother, Verlag von K. F. Koehler. Leipzig. 1934. pp 187, 27 illustrations.

Mein Fliegerleben. Ernst Udet, In Verlag Ullstein. Berlin. 1935. pp 177, 78 illustrations, and Udet's victory list.

In der Luft unbsiegt, Erlebnisse in Weltkrieg, erzählt von Luftkampfern. Edited by Georg Paul Newmann, Major a. D. der Fliegertruppe. J. F. Lehmanns Verlag. München. 1923. pp 302, pp 14 photographs. A series of short essays by the flyers themselves who flew all types of assignments on all fronts.

Unter Stahlhelm und fliegerhaube. Gustav Praclik. Verlag von J. G. Oncken. Kassel. 1936. pp 175, pp 8, photographs. Author flew with FFA 9.

Jadgflieger in Feuer. K. F. Kurt Jentsch, Karl Jofef Sander Verlag. Magdeburg. 1937. pp 254, 16 photographs and the victory lists of the pilots of Jagdstaffel Boelcke totaling 336 and a list of members killed in action. Author flew with FFA 234 and Jagdstaffel 1.

Rudolf Berthold, Ludwig F. Gengler. Schlieffen Verlag. Berlin. 1934. pp 170, 20 photographs.

Kreuz wider Kokarde, Leutnant Ernst Udet. Verlag Gustav Braunbeck. Berlin. 1918. pp 183, 12 photographs and numerous in text drawings possibly by Udet.

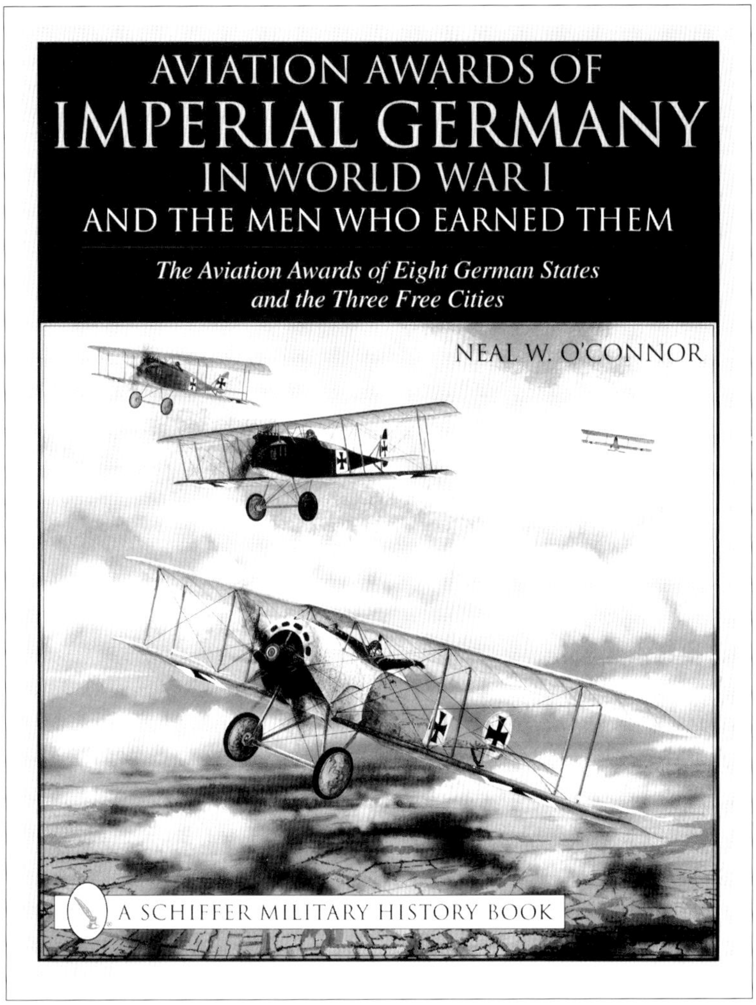

Aviation Awards of Imperial Grmany in World War I and the Men Who Earned Them • *Volume VII*

The Aviation Awards of Eight German States and the Three Free Cities

Neal W. O'Connor

The last of Neal O'Connor's series covering the air awards and pilots of Imperial Germany. The eight German states appear first in alphabetical order: thus we find the Duchy of Brunswick leading off the discussion; the Grand Duchy of Hesse; the Principality of Hohenzollern; the Principalities of Lippe and Schaumburg-Lippe; the Grand Duchies of Mecklenburg-Schwerin and Mecklenburg-Strelitz; and the Principality of Waldeck; and finally, the three Free Hanse Cities of Bremen, Hamburg and Lübeck. Among the forty-three biographies included in this volume are: *Oblt.* Ernst *Freiherr* von Althaus, *Lt.* Erwin Böhme, *Oblt.* Maximilian von Cossel, *Lt.* Ludwig Fritz Dornheim, *Lt.* Emil Flörke, *Offz.*-Stv. Ludwig Friedewald, *Rittm.* Hans-Achim von Grone, *Major* Wilhelm Haehnelt, *Lt.* Franz von Kerssenbrock, *Lt.* Günther Ludeke, *Lt.* Friedrich Niederhöfer, *Lt.* Otto Parschau, *Rittm.* Manfred *Freiherr* von Richthofen, *Lt.* Karl-Emil Schäfer, *Lt.* Carl-August von Schoenebeck, *Lt.* Günther Schroth, *Lt.* Otto-Martin *Graf* von Schwerin, *Offz.-Stv.* Gottlieb Vothknecht, *Lt.* Friedrich-Wilhelm Wichard, and *Oblt.* Heinz-Hellmuth von Wühlisch.

Size: 8 1/2" x 11" • over 700 black and white photographs • 528 pp.

ISBN: 0-7643-1626-5 • hard cover • $69.95

rer erfolgreichsten
ampfflieger.
Vizefeldwebel Festner
Leutnant Schäfer
Leutnan